FRUIT HARVESTING

POCKET NATURE

Discover Nature's Abundance

By Sara Bir

Cover Art and Interior Illustrations by Liana Jegers

POCKET NATURE

FRUIT HARVESTING

CHRONICLE BOOKS
SAN FRANCISCO

POCKET NATURE SERIES

Library of Congress Cataloging-in-Publication Data available.

ISBN 978-1-7972-3764-0

Manufactured in China.

MIX
Paper | Supporting responsible forestry
FSC™ C136333

Series concept by **CLAIRE GILHULY**.

Series design by **LIZZIE VAUGHAN**.

Typesetting by **NGHI TO**.

Cover art by **LIANA JEGERS**.

Interior illustrations by **LIANA JEGERS**.

Typeset in Albra, Benton Sans, Caslon.

10 9 8 7 6 5 4 3 2 1

Chronicle Books LLC
680 Second Street
San Francisco, California 94107
www.chroniclebooks.com

CONTENTS

INTRODUCTION 9

I.
What We Talk About When We Talk About Fruit 17

II.
Where and When to Look for Fruit 41

III.
What to Do with Your Harvest 91

IV.
Fruit and Community 119

RESOURCES 127

If you want to EAT, first you need to learn how to WALK.

—Jonathan Nunn, *Vittles*

INTRODUCTION

Did you know that you can use fruit to tell time? Not in the sense that a clock tells time, with precise minutes and seconds, but in an organic sense, with stages and cycles that indicate what's happening in the larger world. There's the familiar sequence of budding, flowering, pollination, fruiting, and ripening. Each fruit's appearance on stems and branches punctuates the seasons to help situate you in the larger expanse of the world. Months and months later, it all happens again. Fruit is something to savor as a food, yes, but also as a lesson in the importance of following the natural order of things, a reminder that good things can't be rushed.

A fruit hunter uses the natural world as their calendar, and in looking beyond pages of months marked with numbered boxes, they find a reward that spans outward in all directions. Fruit is only a starting point—and a blithely simple one at that. You've bought it at the store in plastic clamshell containers, packed it in your lunch, perhaps learned its names in other languages: manzana, pomme, apfel. Fruit has been with you your entire life; by harvesting it, it can be yours in the most elemental way of all.

Of all pursuits celebrating nature, harvesting fruit is the most immersive and immediate. It rewards every sense: smell, touch, sound, sight, and taste. In peak moments, you can eat it right there, on the spot. It can be as elemental as plucking a velvety warm peach from a tree and sinking your teeth into its yielding flesh, the sticky juices running in rivulets down your arm. That's it: one peach, one tree, one person, one communion. Even that one small act can recalibrate the tenor of your day.

So distanced are we from the source of our food that harvesting fruit yourself—so obvious, so uncomplicated—can feel a bit subversive, maybe even a little wrong. Many times I've shoved berries in my mouth under a tree near a sidewalk and had passersby halt in befuddlement. "Can you eat those things?" they'll ask. I've learned that this type of interaction is not a confrontation but an invitation. It means that they too are curious about this tree and its unfamiliar fruit.

Consider this book to be your invitation. With it, you'll be able to gradually develop a fruit-harvesting pattern that's uniquely yours, specific to your routes and surroundings. Your practice of fruit seeking will continue to unfold with every outing, naturally leading you to be a student of all elements: weather, soil, seasons, and wildlife.

Sometimes fruit refuses to grow despite the most focused and well-informed human interventions; other times, it thrives utterly of its own accord. Even the most ardent fruit hunter will inevitably fail to meet their harvest goals from time to time.

But in my experience, I have discovered that coming home empty-handed does not mean coming up short. My first outings to find fruit were focused on just that—the fruit, something to possess and consume. When I began to forage for fruit, coming home with nothing to show for my time spent searching was the ultimate lesson in patience, which did more for my soul than any high-yielding harvest could have.

Cooking is my trade, and so it was inevitable that the scavenging mindset would eventually include windfalls of free food—the more unusual the better. I hauled home the cardboard box of glossy persimmons that my coworker wearily unloaded to share, then methodically baked persimmon cakes and puddings. While walking my dog, I noticed the quince trees in my neighbor's yard and surreptitiously snatched one or two every day until I worked up the nerve to knock on their door and ask if it was OK. "Is that what those things are?" she said. A week later, I returned with a jar of quince jelly for her.

I kept expanding my lexicon of fruit. It gave a new purpose to the long, meandering stroller walks I did with my infant daughter, snaking through residential streets from shabby, modest houses to the sprawling, landscaped lots in more affluent areas on the other side of the main avenue. My life as a working mother felt small and confined. Noticing the fruiting plants that grew close to my home widened that world considerably. I had a new intimacy with the everyday world that I could access on foot. I could spot fig trees from blocks away and knew from the blotchy rouged spots on a sidewalk when we were coming up on a yard with a purple leaf plum tree.

I'm not a botanist, orchardist, or naturalist. I'm a cook who really likes fruit, and over the years I've learned through repetition where and when to look for it, how to identify and harvest it, and how to harness its power once I get it home. The deeper I get into my fruit-gathering practice, the less I realize I know. It's the best kind of humbling, rife with promise, because there's no end to the knowledge to be gathered.

An unfamiliar fruit is a challenge. How do you get at it? How do you store it? What do you do with it? Fruit can nudge you out of your comfort zone, but never so much that it's intimidating or stressful. In matters of fruit, nothing is life or death. It's just out there, growing, and you either pick it or you don't.

On second thought, fruit is a matter of life or death if you don't know how to identify it correctly, or if you harvest it in the wrong manner, at the wrong time, or in the wrong place. The following pages will help you navigate the ins and outs of fruit harvesting so that you can be a confident, safe, and humane practitioner.

My ongoing quest for fruit has helped shape me into a person I enjoy being. After tromping around scouting for black cherries or mulberries or pawpaws, seemingly urgent dilemmas about work or finances or relationships evaporate, and I realize I'm just another living thing surviving in a world of billions of other living things. I can navigate a new landscape, urban or wild, with confidence, and I will never lack a sense of wonder when passing a familiar tree for the hundredth time.

Harvesting fruit is a thread that never breaks; it doesn't end when the fruit peters out for the season. Every journey, even the shortest or most mundane, becomes an opportunity to observe and notice. A haul of an ephemeral delight can feel like a trophy, but the true spoils are the constantly growing connections with the sky, sun, earth, flora, fauna, and people that you share this world with. To walk outside and know from the smell hanging in the air that a fruiting plant is up to something. To open your front door and relish the sense of opportunity. To allow time to move in unmeasurable moments with no hands or numbers, only instinct.

CITRON, POMEGRANATE,
APRICOT, and PEACH,
Flutter of APPLE-blows
Whiter than the snow,
Filling the silence
With their LEAFY speech,
BUDDING and BLOOMING
Down row after row.

—**Ina Coolbrith,** "Sufficient"

I.

WHAT WE TALK ABOUT WHEN WE TALK ABOUT FRUIT

Welcome to the new, fruit-filled chapter of your life. It is a chapter in which every journey through your front door holds the promise of encountering fruits both familiar and new. It is a mode of being where plants and trees move from the background to the fore, each offering the potential of something enticing to look at, touch, harvest, and maybe even eat. Simply going about your day will offer unexpected sights and sensations that at one time would have been outside your purview. To consider fruit is to enter a dimension of discovery at every turn. All you need to do is linger a little longer and notice what you may once have passed by.

Fruit, it turns out, grows everywhere, which is perhaps why it's so easy to take it for granted. So easy, in fact, that it's surprisingly difficult to think of fruit in an objective sense. What would you say if someone were to ask you, "What is fruit?" Any number of answers would work:

- Fruit is something good to eat.
- Fruit grows on plants and has seeds.
- Plants make fruit in order to reproduce.
- Fruit is sweet, while vegetables are not.

However, all these answers fail to convey the exquisiteness of fruit, how it is both commonplace yet exclusive. Fruit is fleeting. It is often shapely or glossy or brightly colored. It is fleshy and juicy and enticing to eat. Fruit is innocent and indulgent. Fruit is the culmination of a plant's life cycle, its alpha and omega all in one.

Fruit is an elemental part of our diets. "Nana," a baby's babble for "banana," is often one of a young child's first spoken words. It's a testimony to evolutionary magic that our hands are the perfect size for grasping an apple or a banana. Yet somehow, fruit also retains an air of luxury. It's an everyday global food commodity that still feels like a treat. We can sink our teeth right into many fruits without much preparation at all, perhaps save for peeling. They are self-contained packages of perfection, traditionally playing symbolic roles in significant celebrations for nearly every culture: strawberries at Sweden's Midsommar; citrus at Lunar New Year; oranges in stockings at Christmas; dates during Ramadan; pomegranates at Yalda, the Persian celebration of the winter solstice. Fruits play prominent roles in myths and creation stories. To be a living thing means living with fruit.

Botanically, what makes fruit fruit is a mature, ripened ovary, the part of a plant's flower that encloses the ovules. Once fertilized, ovules turn into seeds. The fruit spreads the seeds either by splitting open and/or attracting another living thing to spread the seeds for it.

Under the above definition falls a lot of what we nonbotanists would never think of as fruits: acorns, sunflower seeds, vanilla beans, corn, black peppercorns, summer and winter squash. Plant anatomy, rather than culinary usage, is what makes these things fruit.

For the purposes of this book, we'll mostly go with the culinary definition of fruit, which takes not only edibility into account but sweetness. Beets, which are doubtlessly root vegetables, are sweet, but they are not fruit. Rhubarb, which is tart and decidedly not sweet, sometimes gets lumped in with fruit recipes in cookbooks, but the part that we eat is the plant's stalks, and therefore not a fruit. Tomatoes are botanically a fruit and can be sweet, but we don't eat them as such. You can see that hard-and-fast rules elude us here.

In fact, some of the fruits discussed in this book are not edible—at least for people—but they are attractive or interesting or useful in other ways. Those who become passionate about fruit harvesting eventually become drawn to esoteric or inedible fruit simply because it is fruit and offers qualities that don't have to be consumed to be appreciated.

ACTIVITY

THE SYMBOLIC POWER OF FRUIT

Sweetness is a signifier of caloric density, and as a holdover from our hunter-gatherer days, we are compelled to eat sweet things.

A taste of sweetness can quicken the pulse and trigger cravings for more. Fruit would have been the primary source of sweetness for early humans, and though some fruits lend themselves to preservation through drying or fermenting, it is largely ephemeral.

So it's no wonder that fruit—sweet, fleeting, eye-catching, attractive to people and animals alike—holds a symbolic power beyond its role as a food source. Various fruits have symbolized fertility, luck, new beginnings, abundance, temptation, longevity, wealth, and, occasionally, doom. Think of Snow White's apple, Persephone's pomegranate, and the ancient Egyptian goddess Hathor appearing from the branches of a fig tree to welcome pharaohs to heaven.

Take a moment to reflect on the role specific fruits have played in your personal cosmology. Did you have annual family outings to orchards or go hiking to retrieve berries? Did a backyard fruit tree spawn jam-making weekends every fall? Was encountering a perfectly ripe mango ever a cause for celebration? What bonds did sharing these fruits together create for your family or community? Are there any traditions you'd like to revive or begin? How might harvesting fruit add a new dimension of meaning to these traditions?

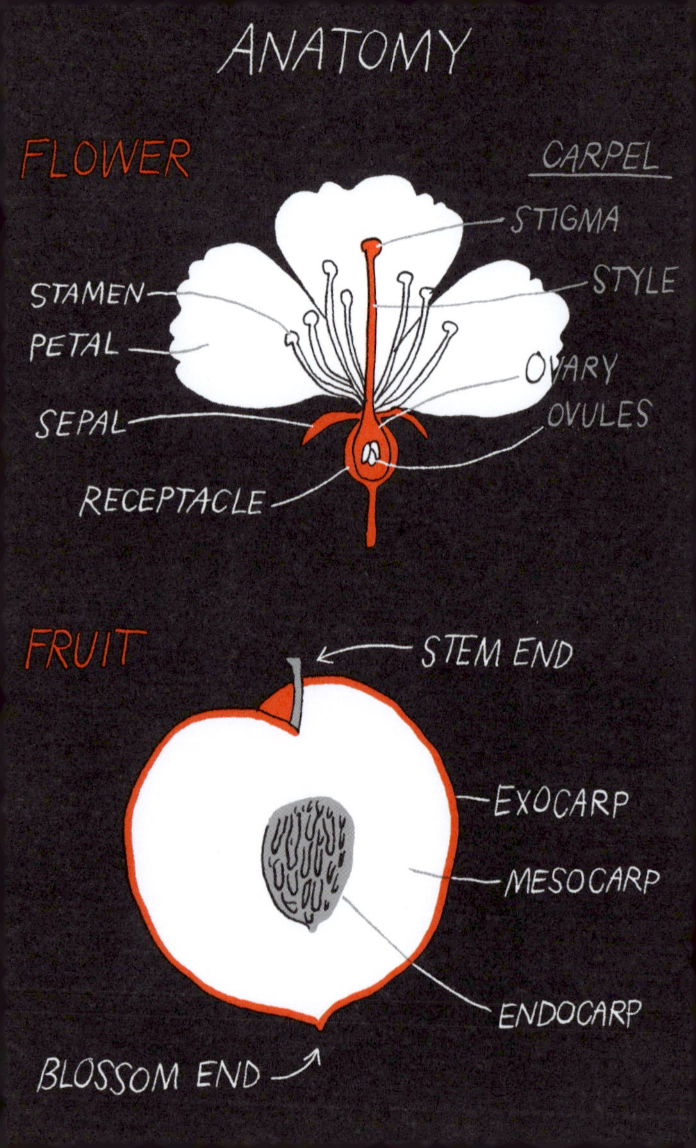
ANATOMY
FLOWER
CARPEL
STIGMA
STYLE
STAMEN
PETAL
OVARY
OVULES
SEPAL
RECEPTACLE
FRUIT
STEM END
EXOCARP
MESOCARP
ENDOCARP
BLOSSOM END

FRUIT AND FLOWER ANATOMY GLOSSARY

Many plant parts play a role in flowers becoming fruit. The culmination of their duties is one or more seeds enveloped in enticing flesh to create a package that will travel away and spark a new generation. When bisected, the fruit's structure bears traces of its previous incarnation as a flower.

- **Pistil:** The collective term for a flower's female parts, including the stigma, style, and ovary.
- **Stigma:** The part of the pistil that receives the pollen. Often sticky, the stigma is where the pollen germinates.
- **Style:** A tubelike extension of the ovary that connects to the stigma.
- **Ovary:** The part of the flower that contains the carpels.
- **Ovule:** An egg. Ovules are found in the ovary and, when fertilized, become seeds.

- **Carpel**: An ovule-bearing structure. A fruit can have one or more carpels.
- **Receptacle:** The end of a flower stalk.
- **Endocarp**: The coating of a seed.
- **Exocarp**: The outer fruit wall.
- **Pericarp**: The fruit wall.
- **Blossom end**: The part of the fruit opposite the stem. This is where the blossom originally formed.
- **Stem end**: The part of the fruit that contains (or contained) the stem.

HOW FRUIT IS FORMED

Fruit begins with flowers, which contain the plant's reproductive parts. In some plants, the flowers include both the male and female parts, while in other plants, the male and female parts exist on separate flowers. The male part of the flower is the stamen, which produces the pollen. The female part, the pistil, contains the ovary at its base, topped by a style and capped with a stigma. The ovary holds one or more carpels, which contain ovules, the "eggs" of the plant. Seeds form when pollen is introduced to the ovules via the sticky stigma. The pollen then germinates and travels down the style to fertilize the eggs/seeds. As the seeds mature, the ovary swells until—voilà!—it eventually becomes fruit.

Plants are said to be "self-fertile" when they don't need another plant to reproduce. This is of concern to fruit harvesters because plants that cannot reproduce from their own pollen are self-unfruitful. "Is this plant self-fertile?" is a common question at nurseries and plant sales among those who hope for fruit, because if a plant is self-unfruitful, it's wise to buy two. If it's self-fertile, you can get away with buying only one. Even plants with so-called perfect flowers, which have both male and female parts, aren't necessarily self-fertile. They may still require pollen from a related plant.

Taking it one step further, dioecious plants have male and female flowers on separate plants. If you want dioecious plants to fruit, you need to plant not only two, but one female and one male. This was the case with the American persimmon seedlings that I bought. The man who grew them couldn't tell me which seedlings were male or female, so I bought three just to increase the odds.

Pollen reaches plants in multiple ways. Bees are the best known, but insects of all kinds do a lion's share of pollination. Wind, monkeys, and bats are pollinators too. Humans can also be actors when they want to ensure pollination. Think of the exacting, hopeful horticulturalist fussing about blossoms with a feather.

FRUIT STRUCTURES

Fruit is not monolithic in its manifestations. In fact, the forms that fruits take are rarely as straightforward as you'd think. We call some things berries when they aren't berries, while other fruits that are berries exhibit few characteristics that laypeople would think of as berryish. Here is a highly simplified guide to classifications that botanists use for different forms of fruit. Understanding the intricacies of a fruit's structure certainly isn't necessary for plant identification, but it can help foster an even deeper appreciation of its marvels. Still, the more you understand fruit, the less you realize you know about it. This list isn't even close to capturing all the different terms used to describe fruit structures; we're just touching on the ones you'll see in culinary fruits, the ones we commonly eat.

▶ SIMPLE FRUIT

A simple fruit develops from a single ovary. Grapes, peaches, and mangoes are all simple fruits.

▶ MULTIPLE FRUIT

A multiple fruit develops from more than one flower—that is, from flowers that grow together in a cluster, or inflorescence (a group of flowers). Examples include figs, mulberries, Osage oranges, and pineapples.

BERRIES

A berry is a simple fruit with a fleshy pericarp. It may contain one or more seeds, and the seeds' endocarp is typically soft. A lot of the fruits we call berries (such as raspberries and blackberries) are botanically not berries, and some of the fruits we don't think of as berries (like bananas and tomatoes) are, indeed, berries. Blueberries, grapes, currants, and cranberries are other examples of berries.

CITRUS FRUITS

Citrus fruits are, in fact, a subset of berries called hesperidia. A thick rind, or peel, is a hallmark of hesperidia. Inside, they have juice sacs separated by partitions that create their segments. Meyer lemons and key limes are examples.

▶ DRUPES

A drupe is a simple fruit with one pit, or seed, in its center. Stone fruits, members of the *Prunus* genus, fit this category. The exocarp is a thin skin; the mesocarp is the fleshy part; and the hard exterior of the pit is the endocarp. Inside the endocarp is the seed, or what we think of as the kernel. Examples include plums, mangoes, autumn olives, olives, almonds, and spicebush.

▶ POMES

Pomes are pseudocarps, simple fruits whose flesh is the developed receptacle rather than the ovary itself. The core of the fruit is its ovary, which contains multiple seeds. Many, but not all, of the members of the Rosaceae family are pomes, including apples, medlars, serviceberries, and loquats.

▶ ACCESSORY FRUITS

The flesh itself of an accessory fruit is studded with "seeds" that are themselves individual fruits. Figs and strawberries are accessory fruits.

SOME FRUIT STRUCTURES

SIMPLE FRUIT

DEVELOPS FROM SINGLE OVARY

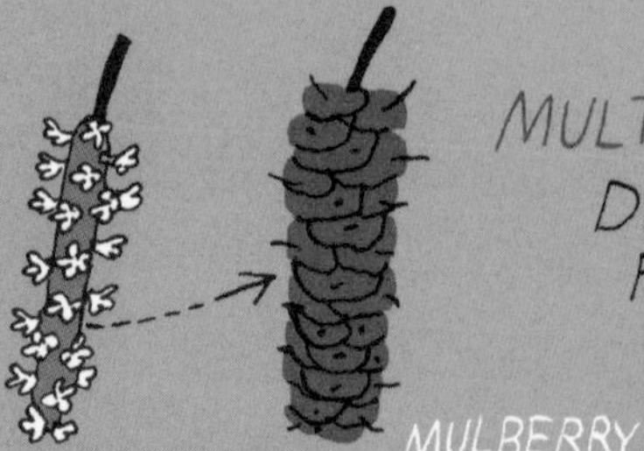

MULTIPLE FRUIT

DEVELOPS FROM FLOWERS IN A CLUSTER

BERRY

FLESHY PERICARP

DRUPE

ONE SEED IN ITS CENTER

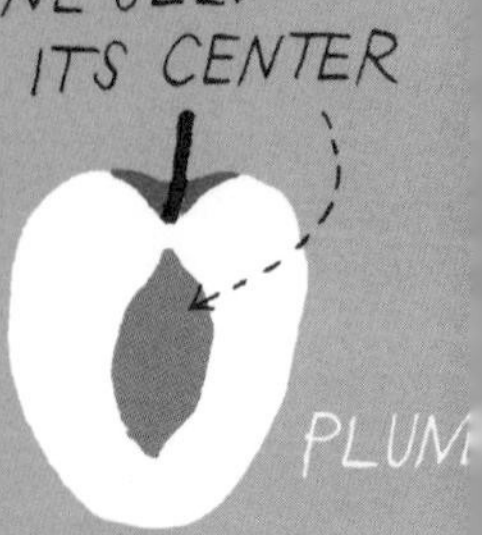

ACCESSORY FRUIT

FLESH STUDDED WITH SEEDS

STRAWBERRY

AGGREGATE FRUIT

A SINGLE FLOWER PRODUCES A GROUP OF FRUITS IN A SINGLE STRUCTURE

RASPBERRY

POME

SIMPLE FRUIT WHOSE FLESH IS DEVELOPED AROUND THE OVARY

APPLE

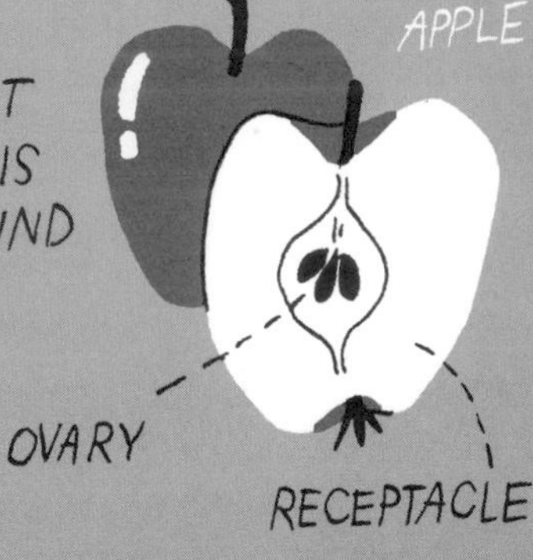

AGGREGATE FRUITS

A group of fruits, produced by a single flower and producing a single structure, is an aggregate fruit. Aggregate fruits can take many forms, including compound drupes (collections of many smaller, one-seeded druplets) like blackberries and raspberries. Strawberries are aggregate accessory fruits where the receptacle enlarges to become fleshy; what we think of as the seeds on its surface are achenes developed from individual pistils. Other aggregate fruits are cherimoyas and jackfruit.

FLESHY SEED CONES

A few outlier members of the fruit family technically aren't fruits at all, since their ovules aren't in ovaries, but are exposed. One such example is a gymnosperm, which produces a fleshy seed cone. Other examples are yew berries, gingko fruits, and juniper berries.

ACTIVITY

SLOW-MOTION PLANT IDENTIFICATION

It's possible to head out the door with only a smartphone in hand and immediately begin identifying plant after plant, even with little knowledge to go on.

Snap a photo and an app like iNaturalist can tell you its species, show you other photos, and tip you off on other locations where you might find the plant. iNaturalist has activated all kinds of burgeoning plant nerds.

There are drawbacks to this near-instantaneous flood of information, however. First of all, iNaturalist works best when you are able to include a clear image of the plant's flower, and we all know that plants are only in flower for part of the year. Secondly, it deprives the user of exercising their powers of deduction.

When I first began noticing plants and foraging, I never left my house without my phone. I'd use photos to help me recall what plants I saw where, in part so I could go back and use the photo as reference when I was hoping to identify the plant with my books. And often I listened to podcasts while I walked.

Eventually, though, I started leaving my phone at home. I wanted to hear the birds, insects, and wind, the leaves crunching under my feet, or the drone of traffic passing on nearby roads. And I wanted to consider the plants without distraction; opening my phone could release a torrent of notifications reminding me of exactly what I was trying to get a respite from with my nature time.

When you're out plant spotting on your own, the journey to identify a strange new plant can be long, but instead of rushing it, consider drawing it out on purpose, simply because you can. It was not so long ago that naturalists, whether experts or amateurs, had to make sketches if they wanted an image of what they'd seen. And they did not always have access to books, but they perhaps had friends, family, and colleagues who could share their wisdom. Compared to today's amateur naturalists, they often had a more fleshed-out library of plant literacy to begin with, because spending time outdoors walking from place to place created a foundation of shared knowledge passed from generation to generation.

You have your own modern-day community and your own problem-solving skills. Take the time to look at a plant closely, multiple times over multiple days, and see how deeply you can imprint it on your memory. If in a public space, collect a few leaves or fruits to bring home as a reference instead of a photo. Borrow a few plant guides from the library, and use them to help narrow down the possibilities. Or go back another day, this time with your phone, and send a photo to a friend who might help you figure out the plant mystery.

I still think fondly of those days when my mind digested stories and interviews through headphones while I roamed sidewalks and pathways in dappled sunlight. I'd scroll through my photo library and relive exciting discoveries from foraging walks. Sometimes I'd arrange the photos and post them as Stories on Instagram, creating a narrative memory of images that captured the thoughts and sensations I had that day. But now that I'm years into my fruit-harvesting practice, I rarely take photos of plants anymore. My relationship with plants is now well seasoned, and I'm content to experience my time with them without dilution.

There are increasingly few times when we are able to linger in a zone of inconclusion; our digitally powered world often requires us to act quickly and often. But those plants aren't going anywhere. There's no rush. Try identifying a plant by degrees—in slow motion, if you will. The sensory aspect will impress upon your brain more deeply, laying a strong groundwork to support your future efforts every single time you meet an unfamiliar new plant. And you'll have the satisfaction of solving the riddle hands-on. We have a multitude of tools at our fingertips; there's no rule that you always have to use the fastest ones. The rewards of information gleaned in this manner gratify on a deeper level when time affords the privilege of pulling it off.

NATURE has never been evicted from cities, only TRANSFORMED.

—**Antonia Malchik,** “A Long Walk Can Change Everything”

II.

WHERE AND WHEN TO LOOK FOR FRUIT

Fruit grows outside, so to find and harvest fruit, you need to intentionally spend time outdoors, exploring new places and revisiting familiar ones with a new, fruit-focused lens.

Plants grow almost everywhere that people are. And many plants have fruit. You don't have to bother with getting yourself to an idyllic setting away from the hustle and bustle. Just go to where plants grow—which is practically everywhere—and in interacting with them more closely, you'll enter micro-pockets of wilderness and placidity. Some of my favorite tucked-away spots for finding interesting fruiting plants are behind storage units, at the

NO
PARKING
IN
ALLEY

far reaches of highway rest stops, and weedy alleys wedged between older residential streets. There's nothing rarefied about such places, but they offer easily accessible portals into resilient plant life that refuses to be tamed.

Start small. Pick a fruit that is familiar or recognizable to you and go for it. If this means apples at a u-pick orchard, it still counts! Focusing on one fruit will reinforce pattern recognition habits that transition effortlessly into the next fruit you add to your repertoire.

If you don't know which fruit to target, begin by noticing a fruit that's commonplace where you are and really dial into it. Consider crab apple trees, which are ubiquitous in yards and all sorts of public spaces. Their fruits are often on their branches year-round, so any season is a good time to begin. Yet there are hundreds of varieties of crab apple trees, and some look drastically different from others. Some have cheerful, cherry-red fruits, while others have mottled gray-green fruits. It'll keep you on your toes, leading you to note the tree's other parts: its leaves, its bark, and

WAYS TO GET FRUIT

There are four popular ways of obtaining fruit. Although these categories overlap, don't think of them as having a hierarchy. In the end, they're all facets of an immensely gratifying pursuit.

HARVESTING • means just that, picking or gathering fruit from a plant.

GLEANING • is gathering cultivated fruit that has been left behind, has fallen, or otherwise won't be used. The term originally applied to gathering the grain remaining in fields after a harvest. Some might also consider reclaiming blemished or excess produce from a market to be gleaning. You might not technically be picking it, but in both cases it's being diverted from rotting and going to waste.

FORAGING • means gathering wild plants. They don't even have to be edible. Fruits can be used for dyes, inks, and floral design. Does harvesting pomegranates from an ornamental neighborhood tree count as foraging? I say yes, even though an ornamental tree growing in a yard isn't wild, per se. Foraging is a mindset as much as it is an activity.

It's moving through the world with an opportunistic lens. To a forager, a plant is never merely a plant. It's an opportunity, an invitation to engage with a living thing and consider what you have to offer each other. What might this plant do for you? What might you do for it? Thus, a cultivated neighborhood tree laden with unclaimed pomegranates becomes an opportunity. That tree could have a story; those fruits could have a home. A forager investigates and observes, uncovering a plant's story before determining whether they should lay claim to the fruit.

GROWING • is cultivating plants yourself. If you are not good at keeping plants alive, beware: Almost every forager eventually becomes so smitten with a plant or two that they decide to grow it themselves, just so they can be closer to it. Even if those plants don't bear fruit (or survive at all), growing fruiting plants is still useful for the intimacy it offers with their habits and growing cycle.

the framework of its branches. A glimpse of crab apples can tip you off, but the sum of all the tree's parts will tell you what it is, be there fruit or no fruit.

SEEKING FRUIT MEANS SEEKING PERMISSION

It's best to make no assumptions about whether it's legal to harvest fruit in any given setting. Always get permission to harvest any plants, edible or not. You can do so by knocking on doors, picking up the phone, or referring to city ordinances.

Oftentimes local laws state that any fruit that lands on the sidewalk side of private property is fair game—that is, if a lemon tree's branches dangle over the property line, the lemons are free for the taking. Even so, not every homeowner delights in a stranger jostling their landscaping around. When in doubt, introduce yourself and announce your intentions. You might just come away not only with fruit but with a fruit-harvesting ally or friend.

Rules vary greatly about gathering anything, fruit or otherwise, from public land. Sometimes regulations are posted on a sign, but usually not. In those cases, make a call or visit the appropriate website.

The golden rule of hunting for fruit: If harvesting a fruit makes you a jerk, don't harvest it. You don't want to damage plants, disturb wildlife, or take all of something and leave nothing for others—humans or animals.

ACTIVITY

PLANT GRATITUDE

Fruit is a gift. We are not entitled to it. Simply being able to walk up to a plant and pick fruit to eat is miraculous. A plant produces it in an effort to reproduce; our involvement may or may not help it achieve that goal.

Think of the Shel Silverstein children's classic *The Giving Tree*. In the book, an apple tree provides a little boy with shade, branches to climb, and fruit to eat. The boy grows into an ever-needful adult, and the tree eventually gives him everything, right down to her trunk, so that when he is an old man, all that she can offer him is a stump to sit on quietly, his last remaining wish.

We don't know if plants have a sense of self, so to say that plants are selfless is a bit of a leap. However, before you harvest fruit, ponder what it would look like to ask nature for permission to do so—not through your words, but through your actions. Take a moment to reflect on not only the fruit itself but also the plant that produced it, the soil that supported it, the sunlight that fed it, and the rains that watered it. In your own manner, make your gratitude known to the natural world for its unrelenting generosity.

There's a way to carry this mindset one step further: Instead of only thinking of what plants can do for you, ponder what you can do for plants. Imagine ways that you can reverse human impact—adding by subtraction, if you will. This could mean volunteering to remove invasive plants or simply collecting and removing litter when you happen to see it. If you have goop-coated pawpaw seeds or persimmon seeds left from processing fruit, you can go to disrupted habitats where those trees would grow and fling the mess randomly about to cast seeds. Your expressions of gratitude can be giving plants the space they need to just be plants.

WHERE TO LOOK FOR FRUIT

Where there there are plants, there is fruit. Plants make it their business to grow wherever they can, which is the most fantastic and admirable thing about them. You'll find plants growing through cracks in a parking lot, along steep riverbanks, in clogged-up gutters, and on mounds of humus on the branches of other trees. The criteria isn't "Is it pretty here? Can I thrive?" It's "My seed wound up here, and I was able to germinate."

This opportunistic aspect of plants can likewise make you an opportunist. There's no need to hold out for sylvan Edens or cloistered gardens or primeval fields tucked away in pristine valleys. Those are splendid places to seek fruit, but they're also uncommon and often inaccessible. Humans tend to think of nature as "us" and "them." There's the world that we have constructed, and then there's pockets of the natural world that we have left untouched. But nature sees no borders;

U
PICK
STRAWBERRIES
RASPBERRIES
FLOWERS

nature just is. It's us and everything around us. Nature is you.

Fruit, therefore, could be anywhere. Just moving around in the world means you can look for it. Edible fruit is more likely to be in some spots than others, though. Consider the following places as you embark on your first fruit-seeking journeys:

▶ PARKS

Parks are a wonderful starting point. When I'm planning a trip, one of the first things I do is look up the address where I'll be staying on Google Maps. I zoom out, find the biggest patch of green that's within walking distance, and then zoom in. If it's a park, I'll click through any photos posted to get an idea of what I might find there. If there are trails, all the better!

▶ FARMS

Farms may seem too obvious, though you likely don't have a farm of your own. This is what u-pick farms were made for. It's clearly transactional, relieving you of any worries about obtaining permission, and the edibility of what you're harvesting is not in question. Farms tend to be sprawling, with scrubby or woodsy untended zones, and those can be fruit-harvesting gold mines. I have casual picking rights on the rural property of a few of my friends. I just give them a heads-up when I'm coming and then help myself to sumac, elderberries, pawpaws, persimmons, and whatever else is plentiful. When you can't cultivate fruit, cultivate relationships.

▶ YARDS

Yards, whether your own or someone else's, are easy to get to and rife with the unexpected. If you have a voyeuristic bent, you can peep at trees and shrubs and think about what they tell you about the people who live there. Tramp around your own yard and try to view it as if you're setting foot in it for the first time ever. You may see unassuming and underwhelming (though still edible) fruit such as false strawberries and ornamental barberries, but it's fruit all the same, useful for training your eye and observing growth cycles. Note, though, that if you plan to harvest fruit from someone else's yard, it's always best to ask for permission first.

▶ TRAILS AND PATHWAYS

Trails and pathways usually have verdant strips on one or both sides. I bike or walk the mixed-use bike path in my town almost daily, which affords me glimpses of many different plants as they gradually do their thing. The downside with busy bike trails is that stopping suddenly in your tracks to examine a promising plant impedes traffic flow, so I try to select more leisurely corridors for plant spotting.

▶ ARBORETUMS, BOTANICAL GARDENS, AND PLANT NURSERIES

Arboretums, botanical gardens, and plant nurseries all have plants helpfully labeled. Seeing them in a curated context with their colloquial and scientific names clearly displayed removes all doubts about what that plant might be, which makes these locations great places to study up on plants of all kinds. It's almost like using flash cards. Go to as many of these as you can and, in time, you'll have memorized various fruiting plants, so you'll eventually be able to confidently identify them yourself in the wider world.

▶ CEMETERIES

Cemeteries are bad places to harvest fruit but often excellent places to become intimate with fruiting plants—and plants in general. I used to walk through a cemetery near my house at least once a day, and there I learned to recognize juniper, serviceberries, barberries, black cherries, gingko, and spicebush. And that's just the fruits! There were also shagbark hickories, pignuts, black walnuts, and stumps with oyster mushrooms. Picking or gathering fruit (or any plant) in any capacity within a cemetery is never permitted—and it's disrespectful. In addition, the water in cemeteries isn't potable, so the fruits shouldn't be consumed anyway. However, enjoying the beauty of a cemetery and considering the contrast between vibrant plant life and the resting places of the departed is a very grounding activity. Take time to notice the monuments and reflect on the lives of the people buried under them. Rather than using the cemetery as a playground for your fruit interests, think of it as a pilgrimage to offer respect and honor the legacies of people you haven't met and can only speculate about.

WHAT'S IN A NAME?

Groupings of cultivated or wild fruit can have different names, depending on what grows there and under what conditions.

ORCHARD • A field planted with fruit or nut trees or shrubs.

GROVE • A small group of trees with little to no underbrush. The term applies to both wild and cultivated groups of trees.

STAND • A group of trees of similar size and age in a forest.

FIELD • Open land planted with crops, typically not trees. Strawberries, for instance, would be planted in a field.

VINEYARD LAND • Where grapevines are cultivated.

BOG • Wetlands with acidic soil (conditions that cranberries and bilberries thrive in). Large artificial bogs are created for commercial cranberry cultivation.

HOW TO IDENTIFY UNFAMILIAR PLANTS

Don't eat plants unless you are 100 percent certain what they are. And how are you supposed to know? Employing a multipronged approach is wise; cross-referencing helps eliminate doubts.

Learning in person is best. Nothing beats seeing and touching or even tasting a fruit in real life. Look for classes and guided walks at local parks or preserves. In lieu of that, if you have friends who are knowledgeable about foraging and edible plants, ask them to go on a walk with you. Most of the foraging types I know would be more than happy to indulge such a request. People who are into weird plants can never talk enough about them.

Keep a few plant guidebooks on hand. I prefer ones that are focused on my region and have clear, full-color images. At least one of them should be the most current edition, since scientific names and taxonomies can shift over time. Still, do not discount decades-old guides, which can offer charmingly phrased glimpses into past sensibilities about nature and our role in it. Once again, ask people in the know for their recommendations, as every intrepid forager has their own library of trusty go-to reference books.

Citizen science apps such as iNaturalist can help you identify a plant from a photo you've taken. Users can post where they have found a particular species, which means that you can use its data for your own research to target a specific fruit.

Social media has boosted interest in foraging, but it's also spread misinformation as enthusiastic newbie foragers start posting before they really know what they're talking about. More visual platforms like Instagram and TikTok are ideal for sharing images and footage of plants out in the wild, urban or otherwise. Facebook groups often include foragers with a range of experiences, without casual comments from everyone and their grandmother. You can establish meaningful connections with like-minded fruit harvesters over social media and then use them as a starting point for direct communication or a way to learn about festivals, conferences, and events that you can attend in person.

Some well-known foraging experts offer online courses, which can eliminate a lot of the white noise of social media. But as a courtesy, don't think of popular foraging content creators as your personal plant ID service. When fans send them direct messages asking, "Is this such and such plant?" or "Is this edible?" it consumes more of their time than you might think. It's unfair to burden them with the responsibility of keeping you safe—that's your job. Do your own legwork, and if you still can't make a confident ID of an enticing plant, there's a simple solution: Don't harvest or eat it.

Besides, this gives you all the more reason to cultivate personal relationships with more seasoned fruit harvesters. A friend who knows about such things is a gift, but so is being that person for someone else. I love it when friends send me plant queries. Sometimes I can't tell them anything conclusive, and we set about solving the mystery together.

HOW SCIENTIFIC NAMES WORK

Understanding a plant's scientific name—the one that's in Latin—can help you navigate through online searches, plant databases, and social media hashtags with speed and confidence. Scientific names can also help you understand patterns in how plants look and behave.

The Swedish biologist Carl Linnaeus developed this classification system in the late eighteenth century. By sorting living things into groups, scientific studies became more navigable. (Perhaps in middle school science class you learned the eight major terms of his taxonomic hierarchy, from broadest to the most specific: domain, kingdom, phylum, class, order, family, genus, species.) He used Latin because it was the shared language of European scholars, and as a dead tongue, it wouldn't evolve the way spoken language does.

Linnaeus's method for naming individual organisms is called the binomial system, since each organism has a unique two-word name comprised of a genus followed by a species name. Thus, the dog rose is *Rosa canina*. The specificity of a scientific name ensures that everyone's talking about the same plant. And the deeper you get into identifying fruiting plants, the more you'll realize what an asset that is.

While the Linnaean system enabled scientists the world over to communicate about biology more clearly, it is arbitrary and not set in stone. In fact, from time to time, scientists reclassify an organism as they discover more about it. This is one reason why it's good to have a recently published plant identification guide around.

Since common plant names can vary by region, using a plant's scientific name when researching its identity is wise because it's much more specific and thus increases the chances of a successful ID. For instance, in the eastern US, *pawpaw* refers to *Asimina triloba*, the largest fruit native to North America. But in some other countries, *pawpaw* is a colloquial term for papaya, *Carica papaya*. Every fall during pawpaw season, some unfortunate publication or website publishes a story featuring photos of papayas, not pawpaws, because their image search was for "pawpaw," which brings up mostly photos of *Carica papaya*, and the photo editor didn't think to carefully scrutinize the images for accuracy.

You can find a plant's scientific name by looking up the common name in a plant identification guide, but a faster way is to search online. Let's say you're looking up red elderberry. In the search engine, search for "red elderberry scientific name." Most of the results will be for *Sambucus racemosa*. Now that you have the scientific name, go back and do an image search for *Sambucus racemosa*, which is more likely to include pertinent images than a search for "red elderberry." Scrolling through a variety of image results will give you more to compare your own specimen with. I still cross-reference with a plant identification book whenever possible because multiple sources will always offer you more weird and wonderful facts to glean.

CITRUS
MALUS
MORUS
PHYSALIS
RIBES
PRUNUS
FICUS
SAMBUCUS

COMMON GENERA OF CULINARY FRUITS

If you're a word nerd, it's fun to know the scientific name of a plant. But instead of memorizing every single one, it's easier to learn the genus name of plants you commonly interact with.

- ***Citrus***: Lemons, oranges, limes, and other citrus fruits
- ***Ficus***: Figs
- ***Malus***: Apples, crab apples
- ***Morus***: Mulberries
- ***Passiflora***: Passion fruit, maypops
- ***Physalis***: Ground-cherries (husk cherries)
- ***Prunus***: Cherries, plums, peaches, nectarines, apricots
- ***Pyrus***: Pears
- ***Ribes***: Currants, gooseberries
- ***Rhus***: Sumac
- ***Sambucus***: Elderberries
- ***Vaccinium***: Blueberries, huckleberries, cranberries, bilberries, lingonberries
- ***Vitis***: Grapes

SEASONAL FIELD GUIDE

Plants are plants all year long, not just when they have fruit on them. Despite commonly held belief, fruit harvesting is a year-round practice, even if you are truly only harvesting the actual fruit in certain months or weeks of the year. As there is great regional variation in where fruits grow and when, this is only a general overview of the cycles of fruit harvesting.

SUMMER

Fruit is off the hook in the wild, in yards, in gardens, and on farms. Summer is the time for keeping up with that abundance as one fruit's peak rapidly ushers in another's.

FRUITS OF NOTE

Gooseberries, blueberries, serviceberries, raspberries, blackberries, sumac, ground-cherries, and stone fruit such as apricots, nectarines, and cherries

FALL

Cooling weather and shortening days bring about longer-keeping, hearty, less ephemeral crops such as apples and pears. Fruit-bearing annuals become dry and scraggly as they die back for the season. Deciduous leaves change color, affording a different way for you to identify fruit-bearing trees and shrubs.

FRUITS OF NOTE

Pawpaws, apples, pears, plums, grapes, figs, quince, cranberries, and persimmons

WINTER

As many perennial fruit-bearing plants go dormant, you can learn to identify trees by their bark. You may find shriveled fruits dangling from vines, branches, and stems. These hangers-on serve as markers for places to revisit in the future producing months. Juniper berries and barberries can even be harvested in the winter.

FRUITS OF NOTE

Oranges, barberries, and air-cured crab apples still clinging to the branch

SPRING

Many perennials bloom, while some annuals begin to sprout. If you want to grow annuals like ground-cherries or melons, now is the time to create your planting strategy. The emergence of new life invigorates an intrepid fruit seeker with the promise of fruits to come.

FRUITS OF NOTE

Meyer lemons, kumquats, and strawberries

SPF
50

THINGS YOU'LL NEED TO HARVEST FRUIT

You don't need much to harvest fruit beyond enthusiasm and a good head on your shoulders. Rather than buying gear, a rummage through your shed or garage will likely turn up the necessary supplies. What really matters is making sure it's handy when you need it. I keep a crate in my car at all times with assorted supplies, all of them either broken or falling apart, which is just as it should be. Your gear only needs to be good enough to get the job done. Here's what I suggest you gather:

▶ OLD CLOTHING

Old clothing that covers your arms and legs can protect you should you be out in thorny areas or near poison ivy or tick-infested woods and fields. Long sleeves and pants can cramp your style in the summertime, but if your skin is sensitive, it's likely worth the temporary discomfort. Assume any clothing will get snagged. I have a bad habit of getting drawn into the woods in some new outfit only to find myself in a thicket of multiflora roses, deadlocked with a tangle of thorns sunk into my one decent pair of leggings. Just know that you won't always be appropriately dressed when a windfall of fruit presents itself. Scrapes, scratches, stains, and tears are occupational hazards that I've come to think of as souvenirs.

▶ A HAT

A hat with a broad brim will keep the sun off your face. It will also theoretically keep ticks off your head.

▶ HEAVY GLOVES

Heavy gloves for yard work will protect your hands from branches and thorns should you feel compelled to go after sumac clusters or barberries. In most instances, though, they'll either be unnecessary or will hamper your ability to work with small or delicate fruit.

▶ GARDENING SHEARS

Gardening shears are useful for lopping the clusters off sumac shrubs. They're also indispensable for the occasional branch or bough, since one way to harvest lots of tiny fruits at once, such as autumn olives, is to simply cut off the fruit-laden branch itself and deal with picking off the actual fruit once you arrive home. (Autumn olive is invasive and plentiful, so in this instance, what seems to be butchery is actually a mitzvah.)

▶ BAGS, BUCKETS, AND BINS

Bags, buckets, and bins will all suffice to hold your fruit, but the right container depends on the fruit itself. Sturdy sacks and cardboard boxes are fine for large, resilient fruits like apples and lemons. Delicate berries fare better in shallow, flat, rigid containers. As berry season approaches, start stockpiling the plastic clamshell containers that produce comes in. (Bonus points for reusing forever plastics!)

- **Old Yogurt Containers:** For picking sturdier berries on the go, I reuse old yogurt containers since they're rigid and have lids.

MAKE A BERRY BUCKET

One of the best reused containers for berry picking is an empty milk jug. My mom would make the buckets for blueberry picking when I was a kid. Wash out a plastic gallon-size milk jug, and use scissors to cut the top off, leaving the handle intact. You can thread a cord through the handle and tie it off to wear the bucket around your neck, leaving your hands free, or you can loop your belt through the handle and carry the bucket at your waist.

I can tuck a full one into my backpack and ride my bike without fear of a spill.

- **Improvise:** Sometimes you'll come across excellent fruit unexpectedly and have to improvise a holder. An inverted hat or bike helmet works great. And there will be times when the only thing you can do is eat as much as possible right there on the spot.

▶ A STURDY BACKPACK

A sturdy backpack with a few supplies tucked inside will be at the ready when opportunity strikes. Mine contains a few plastic grocery bags, work gloves, gardening shears, insect repellant, and sunscreen: almost everything I might need for impromptu fruit harvesting. The backpack itself is old, and the zipper is busted. It's filthy on the inside and has an assortment of desiccated seeds and twigs rolling around the bottom. There's no reason for it to be nice, though, because it's only going to get dirtier. The plastic bags are useful if I encounter litter or need extra capacity for a big harvest.

BEWARE OF PHYSICAL HAZARDS

Fruit harvesting isn't inherently dangerous, but an intrepid fruit seeker will be drawn to fruit that is growing on steep slopes or is surrounded by a thicket of something scratchy or toxic. Before you get carried away, think about the following:

▶ INSECTS

Mosquito bites are a nuisance, but hornets, bees, and wasp stings can be painful or dangerous. Be on the lookout for hovering buzzers and their nests in trees or in the ground. And if you plan on scrambling about in the woods or a field rife with tall grass, ticks should very much be on your radar. You can use a chemical called permethrin to treat your shoes and clothing to repel ticks and mosquitoes, but keep in mind that permethrin is only effective for several washings. Every summer I treat a designated outfit for times when I know I'll be in an area that's likely to have ticks.

▶ POISON IVY, OAK, AND SUMAC

I got the worst case of poison oak I've ever had when I was foraging for blackberries on a trail run. The heat of the sun and the thirst I was seeking to quench must have addled my judgment. If one of these rash-giving beasts grows in your region, learning to identify it is as important as learning to identify the fruit you're looking for.

▶ PESTICIDES AND HERBICIDES

Roadsides, train tracks, and industrial areas are often teeming with handsome fruit, but they are also often places where heavy spraying occurs. Consider if the fruit you're targeting might be contaminated.

▶ SLOPES, RIDGES, AND HEIGHTS

Harvesting out-of-reach fruit from untended trees can provoke harebrained maneuvers. Should you find yourself in a precarious situation where you could fall or something could fall on you, ask yourself: Is this fruit worth the risk of my getting hurt? The answer? Never.

FIVE UNDER-THE-RADAR FRUITS TO LOOK FOR

Your own fruit vernacular will be informed by what thrives where you are. There's always a fruit you wish you could access that belongs to a different landscape or climate. Therefore, offering an inclusive list of fruits to harvest is beyond the scope of this book. However, some plants are endemic in diverse latitudes and habitats. The following five fruits are good bets to look for on your fruit-finding journeys.

Sumac (*Rhus* species)

This small tree (or large shrub) thrives in untamed and developed areas alike. It's a member of the Anacardiaceae family, which includes cashews and mangoes as well as poison sumac (a totally different plant, by the way, from the regular sumac we're talking about here). Its fruit grows in bright scarlet clusters of tiny, seedy drupes that ripen in midsummer. Sumac is tart; ground into a powder, it's used as a spice in Middle Eastern cooking. You can steep the

SAMBUCUS CANADENSIS
JUNIPERUS COMMUNIS
ROSA CANINA
AMELANCHIER ALNIFOLIA
RHUS GLABRA

ripe clusters in water and sweeten it lightly for a refreshing, puckery beverage. Smooth, or scarlet, sumac (*Rhus glabra*) is native to all forty-eight states in the contiguous US. Staghorn sumac (*R. typhina*) has a similar appearance, but its fruit and stems are lightly flocked in the manner of the velvet on a deer's antlers. Fragrant sumac (*R. aromatica*) grows in many states and has a low, spreading form and a more berrylike fruit. Sumac generally prefers dry soil and open areas. Only the female plants flower and fruit, and colonies are often single-sexed, so finding lots of sumac trees or shrubs doesn't mean you'll find fruit. In the fall, the foliage of sumac trees turns a bright crimson, making the trees easy to spot.

Juniper (*Juniperus* species)

The whole, dried juniper berries that lend gin its piney punch come from common juniper (*Juniperus communis*), which is aptly named, as it can thrive in many locations, particularly those with dry soil and plenty of exposure to sunlight. Eastern red cedar (*J. virginiana*) can be found in woods and edge habitats in thirty-seven different states. Its branches are especially prickly. In western states at higher elevations, you may find western juniper (*J. occidentalis*) growing in rock outcroppings. Junipers are go-tos in public landscaping and well worth looking out for. Juniper berries (which are actually seed cones) are bluish-purple when mature and can be picked and dried at any time of year, as they stay on the branch after ripening. Make sure to do your research when harvesting, as not all junipers are edible, and even with the ones that are, the berries should be consumed in moderation as they are pungent and can be toxic when overconsumed.

Serviceberries (a.k.a. Saskatoon Berries, Juneberries, and Sarvis) (*Amelanchier* species)
These delightful little members of the rose family suffer from a publicity problem. First of all, they go by too many names. Secondly, too few people know of the fruits. Not many are cultivated commercially; the trees themselves are more commonly grown as street trees.

This is good news for urban foragers because serviceberries are delightful treasures to eat on the spot in the few early summer weeks when you might encounter ripe ones. They're a purplish hue when ripe and have a flavor that's like a blueberry infused with marzipan. Serviceberries are native to North America and can be found in woodlands and mountainous areas, but the ones grown in the wild often pale in comparison to the plumper ones that get full sun.

Elderberries (*Sambucus* species)

Where there's sloshy ditches, ponds, creeksides, and wetlands, you'll perhaps come across a sprawling elderberry shrub. The flowers flavor the delicately perfumed liqueur St-Germain, but their bead-size berries can be enjoyed in cordials, syrups, pies, and jellies and are prized for their immune-boosting properties. Common elderberry (*Sambucus canadensis*) is found across North America, while blue elderberry (*Sambucus nigra ssp. cerulea*) appears in western states. Eat only the berries, and take care to thoroughly pick off the stems, which are toxic. Always cook or ferment elderberries before consuming. Avoid eating them raw (they don't taste good without sugar anyway).

Rose Hips (*Rosa* species)

All roses, once their blooms fade away in the fall, reveal their fruits, which we call rose hips. Their size and harvestability depend on the species; the big, plump, fleshy ones are the best to look for. In rosy thorny fashion, rose hips have stiff and prickly little hairs, so handle with care. Jams, jellies, and syrups are traditional preparations for these sour and tricky fruits. Rose hips can also be dried and used in herbal teas, harnessing their plentiful vitamin C. Dog roses (*Rosa canina*), prairie roses (*R. arkansana*), and beach roses (*R. rugosa*) are all good bets for prime rose hips. For the best flavor, collect them once the weather cools, preferably after a frost.

ACTIVITY

READING THE LANDSCAPE

Revisiting the same sites at regular intervals gives you an integrated understanding of the landscape. (Daily is ideal, but weekly or monthly will do.) Given time, you'll develop "plant literacy," and when you are in a similar setting in a location that's new to you, your brain will automatically seek out the types of plants it knows from previous outings. It'll happen subconsciously, like reading words.

Plant literacy could be thought of as the antidote to plant blindness. The concept of plant blindness was created by the biologist-educators James H. Wandersee and Elisabeth E. Schussler in 1999 to articulate humans' collective unawareness of plants as meaningful and individual components of our world, a green blur that's "mainly a backdrop for animal life." They wrote, "We can only recognize what we know. Psychologists would say that plants have low signal value for most humans today. Once objects have acquired meaning for an observer, they are more likely to bc perceived (inattentional blindness). In other words, inattention becomes attention if a stimulus has meaning." Because we don't need to be closely dialed into plants to survive in today's world, plant literacy is not widely carried forward in generational knowledge and formal education.

That's a severed thread that you can heal in part by training yourself to see plants through your practice of seeking fruit. With enough time, you'll be able to assess unfamiliar surroundings and anticipate what plant life it might hold. Near a creek, pond, or trench? Look for elder trees. Is there a black cherry tree? Likely there will be others nearby. Hunting for pawpaws in the woods? Pay special attention to shady slopes and gullies. See lots of 8 to 12 ft [2.4 to 12.6 m] tall flowering trees in the strip between the sidewalk and street? They could be crab apples, ornamental cherries, or serviceberries. Does a yard offer a glimpse of the distinctive shape of fig leaves? Fig tree enthusiasts usually can't stop at just one fruit tree, so there are likely other trees of note in that yard.

I think of this practice as reading a landscape. My brain now does it automatically, often without my awareness. When I'm visiting a place for the first time, whether it's a city or a state park or a housing tract in a suburb, I look for plants. What I see not only reveals information about the people there but also reassures me and offers guidance for a path ahead. If I can read the landscape, I know what's coming up, and I never feel lost. Familiar plants mean I belong there.

Choose a place that you frequent, and apply this practice. Perhaps on the walk from your bus stop to the office, you can implement the habit of reading the landscape, subtly noting the flora and fauna, day after day, expanding your understanding of the environment around you. Breathe it in and let it gradually become part of you.

Then, the next time you travel somewhere new, take in its unfamiliar landscape and see if you can find touchstones that echo your own home landscape, the one that's inherently part of who you are. Linger on plants that you're familiar with, and consider whether they help situate you. Anticipate what else you may find as you explore: Might you come upon streams, ditches, or ponds? What types of animals or human development might you expect to see, given what you know from prior experience? See what parts of this new landscape you can add to your own sensory library. When you return home, will you experience your home landscape in ways you didn't before?

Said BUSY, BUSTLING Mrs. Quince:
"I never have a moment since
THE JELLY-MAKING TIME IS HERE;
WE'RE MAKING such A LOT this year."

—Elizabeth Gordon, *Mother Earth's Children*

III.

WHAT TO DO WITH YOUR HARVEST

You did it! You harvested your treasure of living gems and can celebrate your spoils. You have a decision to make. Now is the point of demarcation. Either you sink your teeth into it right then and there or haul it home with designs to make it last longer—days, weeks, months, or even years.

EATING FRUIT IN SITU

One of the most visceral thrills in the living world is consuming fruit directly from the branch, its skin glowing with the luminous hues of ripeness, its juices running in rivulets down your wrist to your elbow. The air might be filled with birdsongs and wind rustling through tall grasses. In that moment, human exceptionalism melts away, and you are an animal like all the other animals that feast upon that fruit. The skin or the core or the seeds you can fling away without a care, knowing other seeds and skins will join them on the ground eventually. The best fruit never tastes better than it does in the shade from whence it came.

Of course, some fruits, such as quince, are inedible unless cooked. And some—quince again—benefit from added sweetness. And other fruits grow in amounts ample enough to beg to be carted home. Either way, it's not only the fruit itself that you carry with you, but the echo of its memory.

TO RINSE OR NOT TO RINSE

Every u-pick operation I've visited cautions visitors to wash any harvested produce thoroughly before consuming, yet don't we all delight in treating the operation as if it's a giant fruit buffet with an unspoken "don't ask, don't tell" policy? But of the strawberries I eat throughout the year, only a small fraction go right from the field to my mouth.

I always rinse fruit that comes from a u-pick farm before I eat it; even if it's organic, they may still use pesticides or herbicides that can leave a residue. And fruits of all sorts, especially those growing close to the ground, can harbor dirt and bugs, which shouldn't make you sick but can potentially spoil your appetite.

PRESERVATION METHODS

The gift and occasional curse of fruit harvesting is a harvest that's too large to use up before it rots. This is where ancient and modern preservation methods leap to the rescue! Unsurprisingly, sometimes the ones that require the least energy (both from you and from the power grid) are the most ancient.

The best way to store fruits depends on the fruit in question; the subtleties and best practices are vast. So it's impossible to advise on one way to keep any fruit around for maximum enjoyment that you may drag home. This is where you'll have to do your own legwork. But here are a few old, reliable techniques:

Refrigeration

Some people reflexively put all food in the refrigerator, while others rarely ever refrigerate fruit. Neither camp is necessarily wrong. A fruit bowl on the counter is fine for storing most ripe, sturdy fruit you plan to eat within a few days. A giant harvest of strawberries can sit out in a cool spot for maybe a day. But plenty of fruit benefits from refrigeration. Some apples will keep under refrigeration for months. Notoriously ephemeral pawpaws can

Hoshigak
Quince

Rosehips
Raspberry Jam

have their window of wellness extended for up to a week under refrigeration. Refrigeration can help you bide your time as a holding place: Harvest the fruit one day, collect your resources and supplies, then make jam, bake pies, or toss into a simmer pot.

Freezing

The plus side of freezing is that it's comparatively easy when stacked up against other preservation techniques. The downside is that, for many of us, freezer space is finite. Also, once thawed, frozen fruit tends to be mushy. If you make a ton of smoothies or bake a lot of pies and cobblers, the ability to freeze fruit can be a blessing.

I like to spread berries on a baking sheet, freeze them until firm, and then bag them up. I don't rinse off most berries prior to freezing, as they tend to hold up better in the freezer that way. What I usually do is remove what I need from the bag and rinse it off while it's still frozen.

Most frozen fruit is best used within a year, particularly if your freezer tends to cause freezer burn. But I have used persimmon and pawpaw pulp that was over two years old and didn't notice any loss in quality.

Canning

Compared to other preservation methods, canning is a new kid on the block, despite any quaint homesteader associations we may have with it. (In fact, it was developed by a scientist during the Napoleonic Wars in response to prize money offered for innovative ways to feed troops.) It requires special knowledge and equipment and, for me at least, always involves making a big mess. But I love canning, and every year I put up many pints of peaches, applesauce, and assorted jams. One advantage to canning versus freezing or refrigeration is that canned foods are shelf-stable. They're also charmingly giftable.

Drying/Dehydrating

Some fruits lend themselves to drying better than others. The season and the humidity level have a lot to do with the success you

FRUIT SOUVENIRS

Preserving your fruit harvest offers practical value, yet its most lasting worth might be more about the preservation of the person you were when you harvested it. Dried, canned, or infused fruit concoctions in bottles and jars serve as mementos whose presence can conjure up visions of fruit harvests of yore, long-ago days in perhaps faraway places. I keep a cabinet of vinegars, herbs, liqueurs, bitters, jams, and assorted fruity specimens as a scrapbook of sorts that transports me to treasured journeys and stolen moments: A riverbank in Missoula, Montana. A hidden staircase behind the Shedd Aquarium in Chicago, with Lake Michigan looming on the horizon. A stroll through a quiet neighborhood in Portland, Oregon.

As you preserve your fruit harvest, you can label and date your bags and bottles in any way you see fit. Note the date and contents, of course, but also the location and even the people you were with that day. You should relish each item's consumable contents but also cherish it as evidence of fleeting adventures.

may have dehydrating fruit without the aid of a dehydrator or oven. Dehydration methods differ a great deal from fruit to fruit too. However, to make fruit leather (page 107), you only need an oven, the right mix of fruit, and a good set of instructions. Hoshigaki, or Japanese-style dried persimmons, don't require any special equipment, but they do require periodic gentle massages and an abundance of patience.

Fermenting

Brewing, curing, and pickling fruit opens up a whole new dimension that can be especially pleasing for those who don't do dessert. Many cultures have their own fizzy fermented drinks—think of Mexican tepache—but there's also ways to do secondary fermentations for fruity kombucha or natural sodas. Lacto-fermentation occurs when you salt-cure ume (a Japanese apricot). You can try adding an element of fruit to kimchi or sauerkraut for a balancing sweetness. One of the easiest home fermentation projects for fruit is

fermented berry syrup (page 111). One of the trickiest, surprisingly, is making fruit vinegar from scratch, but it's worth a shot if you have fruit to spare.

Infusing

Steeping fruit in brandy, vodka, or other alcohol to make flavored extracts, liqueurs, and bitters can fortify your bar cart in unique ways. Since I'm terrible at fermenting fruit into vinegar, I sometimes go the lazy route and infuse purchased vinegar with fruit.

SCARCITY AND ABUNDANCE

Leaving fruit to rot on the branch or ground can feel like a squandered opportunity when others could surely benefit from it. Yet is it your duty to find a loving home for every fruit? In my first few seasons of fruit harvesting, I didn't want any fruit to go neglected, and I scrambled all over town in a frenzy, shaking mulberries from trees and swooping up pawpaws that had fallen. But keeping up with it all got to be too much, and I realized that I was creating stress where there was no need for it. Fruit never truly goes to waste. Birds, insects, and mammals will come and gobble it up. What's not eaten will rot and enrich the soil. Your harvesting of fruit is only one of many possible outcomes.

The flip side of abundance is scarcity. Whether cultivated or wild, fruiting plants inevitably have years when they don't produce. It could be because of a late freeze or a drought or an infestation or a disease. It could simply be because that plant is fickle. For a recreational fruit harvester, scarcity is disappointing, but in our world of global food commodities, we're fortunate to be in a situation where the fruit we harvest ourselves is nice to have rather than essential.

alad
lade
Marmal

FRUIT LEATHER

INGREDIENTS

At least 1 pound [455 g] ripe, fleshy fruit of your choice

Sugar, honey, or maple syrup

1 teaspoon lemon juice, lime juice, or apple cider vinegar, optional

As a kid, I enjoyed Fruit Roll-Ups in my lunch, but they likely contained just as much corn syrup as fruit. Homemade fruit leather is much more wholesome, and it's a fine way to make use of mounds of fruit.

This method is adapted from Elise Bauer of the website Simply Recipes. Think of it as a technique rather than a recipe. Your yield will depend on the type of fruit you're using, whether it's raw or cooked, and whether you're including the skins or peels.

INSTRUCTIONS

1 Rinse the fruit. If you're working with apples or thick-skinned pears, you may want to peel the fruit. Peeling stone fruit is up to you. Next, pit or core the fruit if necessary, and chop it.

2 Put the prepared fruit in a large saucepan and add about ½ cup water for every 4 cups of fruit. Bring it to a simmer over medium heat, cover, and cook over low heat until the fruit is soft, about 10 to 15 minutes. Mash the fruit directly in the pan with a potato masher or sturdy spoon. If the fruit is especially seedy, you can run the mash through a food mill or conical sieve to remove the seeds.

3 Taste the mashed fruit and add sugar or other sweetener to taste. If it tastes flat, add a teaspoon or so of lemon or lime juice or apple cider vinegar. Return to a simmer and gently cook, stirring, until the sugar dissolves and the mash thickens to the consistency of thick baby food, about 10 minutes.

4 Puree the fruit in the pot with an immersion blender or in a blender or food processor. Taste again and adjust the levels of sugar and acid to your liking.

5 Line a rimmed baking sheet with parchment paper and pour the puree in an even layer about ⅛ to ¼ inch thick. (If you have a dehydrator, cut the parchment to fit your trays.) Depending on how much mash you have, you may need more than one baking sheet.

6 Heat the oven as low as it will go. If that's above 170°F [75°C], you may want to consider propping the oven door open with the handle of a wooden spoon. If you have a convection setting, use it, but keep the oven door closed. Place the baking sheet in the oven on the center rack and leave it until the fruit is firm and no longer sticky to the touch. It should easily release from the parchment. Depending on your oven and the moisture of your fruit puree, this could take anywhere from 8 to 20 hours.

7 Let the fruit leather cool, then cut it into strips. (I leave it on the parchment paper and cut right through the paper and the leather at the same time with scissors.) Then roll it up with the parchment still on there to keep the fruit leather from sticking to itself. The fruit leather will keep at room temperature in a tightly sealed bag for several months, or you can refrigerate or freeze it for longer storage.

Blueberry
Chokecherry
Serviceberry

FERMENTED BERRY SYRUP

INGREDIENTS

At least 2 cups [280 g] berries, such as elderberries, aronia, cranberries, serviceberries, blueberries, grapes, chokecherries, tiny plums, ground-cherries, gooseberries, or currants

Enough sugar to equal the volume of your berries (e.g., 2 cups [400 g] sugar for 2 cups berries)

Inevitably, those who don't care for pies and preserves hit a wall. Isn't there something a bit more grown-up to do with succulent berries? Yes, there is, and it's also largely hands-off. It does involve a ton of sugar, though.

Berries I've fermented into syrup include elderberries, chokeberries, cranberries, serviceberries, and blueberries. Raspberries and blackberries might be too delicate, but you could try them. Grapes, chokecherries, tiny plums, ground-cherries, gooseberries, and currants might work too.

I've even blended different syrups together the way a winemaker blends wine. You could use these syrups in cocktails, but I just add a little bit to plain fizzy water for a simple yet non-cloying mocktail.

INSTRUCTIONS

1 Don't rinse your berries; the natural yeast on their skin will boost fermentation. Toss the berries and sugar together thoroughly in a large bowl.

2 Transfer the sugared berries into a sanitized glass jar that's roughly double the capacity of the berries that you're fermenting. Screw on the lid.

3 Set the jar on the counter in a place where you'll see it. Gently shake the jar every day to help the sugar dissolve and to prevent mold from forming. Over several days, the berries will release more liquid, and most of the sugar will dissolve. When you see tiny bubbles clustering around the sides of the jar, you'll know the berries have begun to ferment. Some batches are slower to ferment than others; it all depends on the ambient temperature of your kitchen and any natural yeasts that were present on the skin of the berries. Be patient. Holding out for this longer fermentation leads to a more rewardingly complex syrup.

4 As the syrup ferments, taste it. When it's fermented to your liking, strain out the fruit. Bottle the syrup and keep it refrigerated. It will last for up to 4 months, though I've had batches remain good for well over a year.

FORAGED INK

Pokeweed berries (whose seeds are toxic; don't eat them) make a delightful fuchsia ink.

Mulberries and blackberries make a purplish-black ink that's fun to use but fades quickly once it's on paper.

Aronia, simmered in a little water for 10 minutes, makes a thick and sticky dark magenta ink.

Wild purple grapes—the tiny ones that are as hard as pebbles—can be simmered in enough water to cover for about 5 minutes, crushed as they are cooking, and then strained for a mind-blowingly purple ink.

Black cherries and chokecherries, prepared just like the wild grapes, yield a saturated brownish-red ink.

For times when delectably edible fruits are in short supply, you can still enjoy harvesting wild fruit that's not choice eating and make it into ink. It's incredibly simple, but it feels like a magic trick, so accustomed are we to paints and fabrics with manufactured hues. It can be easy to forget that not so long ago, all pigments were naturally derived.

POKEWEED
BLACK CHERRY

The most straightforward inks to make are with deeply hued berries such as mulberries and blackberries, which you can crush and strain. Voilà! Ink. Yes, the process is messy and you don't get a ton of ink, but it's fun. The tone of the pigment varies depending on the pH of the paper you use it on, and some inks fade over time. Think of this unpredictable behavior and slow degeneration of foraged ink as part of its appeal, a way for the fruit to continue telling its story.

INSTRUCTIONS

Refrigerate your berry-based inks, adding a whole clove to the bottle to aid in preservation, and toss them when their pigment dims. There are longer-keeping fruit-based inks, such as buckthorn, pomegranate, and avocado, that require more involved processing and kitchen chemistry; if you become enamored with making ink, you may have to adapt formulas for plant-based dyes, as it's easier to find books on making plant-based dyes than plant-based inks. Firsthand experimentation is ultimately the key with these, so I encourage you to embrace the mess and learn as you go.

ACTIVITY

THE JOYS OF TEDIUM

With many fruits, processing your harvest is repetitive, dull, and slow work. It can take any combination of picking, peeling, seeding, pitting, coring, slicing, pulping, and straining to fashion fruit from something unruly and elemental to something refined.

The sensual pleasures and tactile gratification of handling natural material directly is, for many of us, a much smaller part of our daily lives than it was for our ancestors. If your days are filled with futzing over a spreadsheet or debugging computer programs, working with material tedium can balance out the drain of digital tedium.

As you stem elderberries, pulp pawpaws, or singe glochids from prickly pears, find a flow state through handiwork and physical contact with organic objects. Consider how the skin of

the fruit feels under your fingertips, or listen to the thudding of berries hitting the bottom of a nearly empty bucket. As you move deeper into your task, do your hands move faster or slower? Does the light change at all as minutes creep by?

Observe the slow transformation of your project, whether it's whole fruit to preserves or persimmons to hoshigaki. Later, when you open a jar of jam or a bottle of ink, think of it as opening a time capsule, and recall the actions and sensations that it took to fashion your fruit into something exquisitely new.

You are undone
if you once forget
that the FRUITS OF THE EARTH
belong to us all,
and THE EARTH ITSELF
to nobody.

—**Jean-Jacques Rousseau,** *On the Origin of the Inequality of Mankind, Second Part*

IV.

FRUIT AND COMMUNITY

Fruit harvesting can be a solitary pursuit, but even when you're out solo, you're always connecting to your community.

The nature of fruit is that it's not around, and then, seemingly suddenly, there's a surplus of it. Fruit gives us a reason to share. The bounty transforms the often-private experience of harvesting fruit into the communal experience of distributing it. This could be leaving a box of fruit on a neighbor's doorstep, dropping off a surplus of apples at a food pantry or community meal program, baking a pie and having friends over for coffee, or hosting a jam-making session—or asking someone to show you how to make jam.

FRUIT-HUNTING COMPANIONS

If you live in a region with pronounced seasons, the late fall and winter months can be bleak on the fruit front, and much of the fruit that remains on branches are small red berries that are either toxic or unpalatably bitter. Holly, winterberry, barberry, mountain ash, bittersweet, and hawthorn are all such fruits, their splashes of color lending cheer to a monochrome landscape.

While you may not be harvesting these berries, they're much coveted by birds. Insects are scarce in cold months, and such berries become a valuable food source for our feathered friends. Interestingly enough, birds are instinctually drawn to red, and such plants likely evolved to produce red fruit as a means of attracting birds to eat and then disperse their seeds.

When you can't find fruit to harvest, you can turn your attention to bird-watching. In this manner, birds are your unwitting companions. Knowing that you're not the only one on the prowl for fruit makes solitary walks feel less so.

The most insightful bits of knowledge are often passed directly from person to person. When I'm out foraging and someone approaches me to ask what I'm doing, I like to tell them and I ask if they've had this fruit before. If they answer yes, I ask them where and what they like to do with it. If they say no, I give them the elevator pitch, hoping that I'll make them a convert. Really, though, when people notice your fruit harvesting, they may be looking for a chance to better understand the place where they live and the other people who live there.

Random conversations have led me to a faster way to process pawpaw pulp and taught me not to bother peeling or pulping American persimmons—just removing the seeds is enough, saving time and hassle.

FINDING FRUIT EVENTS

The solace of a solitary fruit hunter is poignant and sacred. Few moments can feel freer than when out seeking fruit alone. The flow of time shifts, both speeding up and drawing out. Your sense of self evaporates away; no one is in charge of you, and no one is there to care what you do.

It is lonely, though. Plants have much to teach us, but as social creatures we need the companionship of other people to exchange ideas and to revel in our appreciation of the same wonderful thing.

Fruit people are the nicest people. They are the least likely to judge you, and the most likely to embrace you. They're also sometimes the weirdest, which is why it's best to go to fruit-centric events with an open mind and few expectations outside of having a good time. Think about where fruit people congregate and go to those places, whether conferences, festivals, plant sales, or u-pick orchards.

There are more festivals focused on specific fruits with local significance than you might guess; multiple festivals honor pawpaws, persimmons, and serviceberries, for instance. Some of these festivals are more about fair-like amusements than the fruit itself, while others are small in scope, perhaps only half a day long and contained to a fraction of a parking lot, offering tastings, information for growers, and opportunities to ask experts questions. If there's a fruit-centric festival near you, by all means go; there's no better place to find your people.

Plant sales, particularly native plant sales, will similarly have many like-minded people all in one spot. They're worth going to even if you don't plan on ever growing plants yourself, because you can see lots of plants all in one location, along with the people who know and love them.

A suspected mutual love for a peculiar fruit can be a wonderful conversation starter. An attraction to unusual fruits bridges religious, political, and educational differences. When you go out seeking fruit, be open to not only the flora and fauna but also the people you may encounter. Engage with them. You can make their day, and vice versa.

ACTIVITY

LOSE YOUR EGO

The concept of self and identity looms large in contemporary society. Our opinions are constantly being solicited, our personal images scrutinized. And while time outdoors immersed in plant life is great for reflecting on your role in the universe, it's also a rare opportunity to not think of yourself at all.

Plants carry on doing their thing with no consideration of what you think, what you look like, or who you are. Your role is to help disperse its seed, and that is all.

The minutes or hours you spend harvesting fruit can take you out of your own head and into the larger world, where your role is but a tiny part of a larger, integrated system. In this context, you are not a unique individual, but a vital component of something inconceivably holistic and complex.

When you journey out to seek fruit, try to gradually turn your mind away from the specific worries and tribulations of your relationships, finances, or work. When you notice your mind wandering to such thoughts, take a moment to look up at the sky or down at the ground and reroute your attention to what is in front of you, not inside your head. Feel the boundaries of your personhood and your surroundings blur as the air, sunlight, and sounds envelop and embrace you. Keep harvesting until you lose your sense of self and simply exist. Afterward, once you've returned to life as usual, pause for a moment and consider if you still feel an echo of your earlier equilibrium with the natural world.

RESOURCES

Bir, Sara. *The Fruit Forager's Companion: Ferments, Desserts, Main Dishes, and More from Your Neighborhood and Beyond.* Chelsea Green Publishing, 2018.

Elpel, Thomas J. *Botany in a Day: The Patterns Method of Plant Identification.* HOPS Press, 2013.

Lady Bird Johnson Wildflower Center, University of Texas at Austin. www.wildflower.org.

Lebo, Kate. *The Book of Difficult Fruit: Arguments for the Tart, Tender, and Unruly (with Recipes).* Farrar, Straus and Giroux, 2021.

PLANTS Database, USDA Natural Resources Conservation Service, US Department of Agriculture. https://plants.usda.gov/.

Various authors. *Wild Berries and Fruits Field Guide.* Adventure Publications, 2011. Series of five regional guidebooks.

Yoest, Helen. *Good Berry, Bad Berry: Who's Edible, Who's Toxic, and How to Tell the Difference.* St. Lynn's Press, 2016.

SARA BIR is a chef, writer, and food librarian. She's the author of *The Fruit Forager's Companion*, *Tasting Ohio*, and *The Pocket Pawpaw Cookbook*.

SEE MORE BOOKS IN THE SERIES

THE POCKET NATURE SERIES offers meditative and insightful guides to reconnecting with the natural world through mindful practices.

WWW.CHRONICLEBOOKS.COM/POCKETNATURE